Stunning Pictures Of Space For The Modern Day Astronomer

By Sir Richard Sullivan The Third

Copyright © 2015 Sir Richard Sullivan The Third

All rights reserved.

ISBN: 10: 1522774521
ISBN-13: 978-1522774525

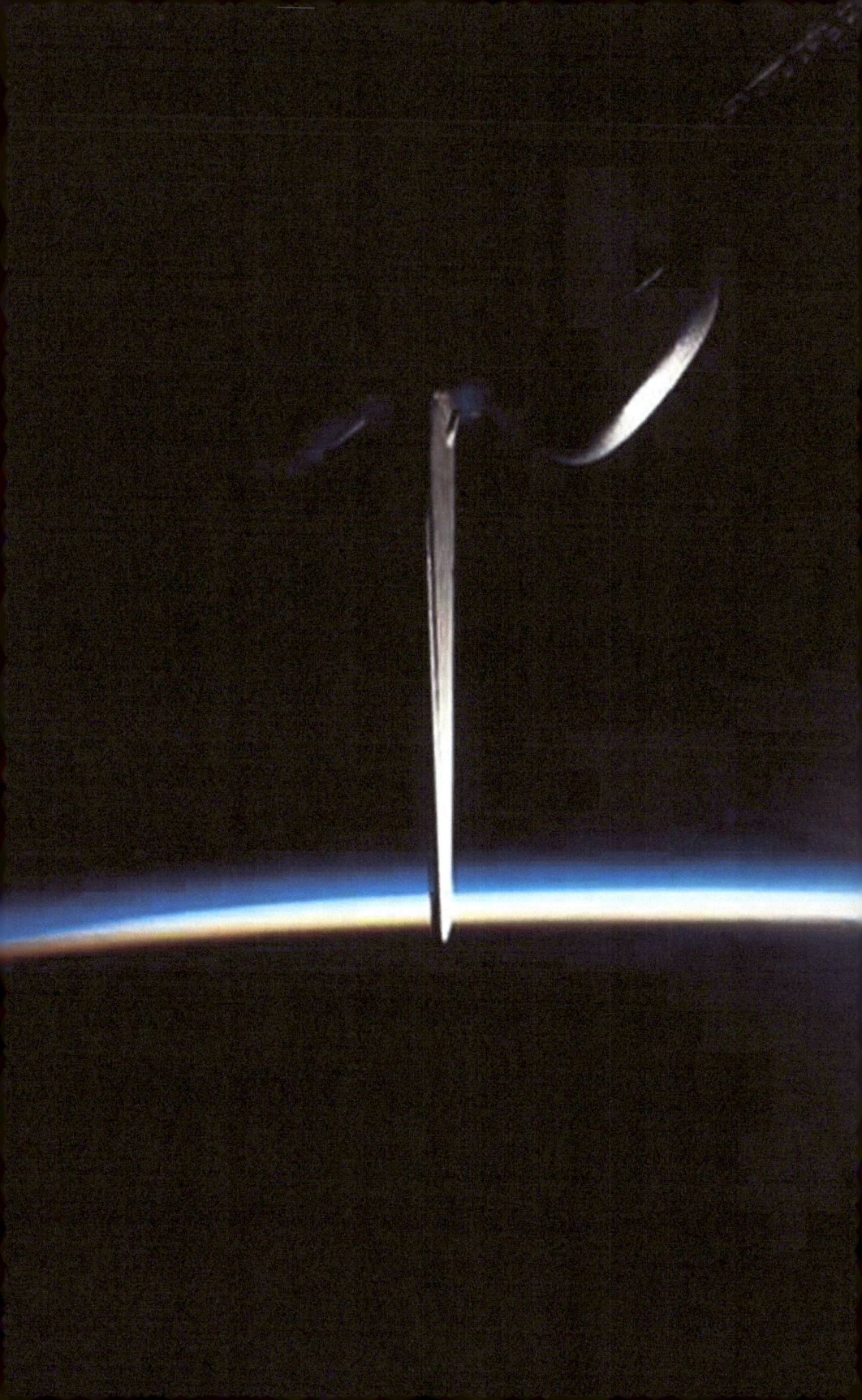

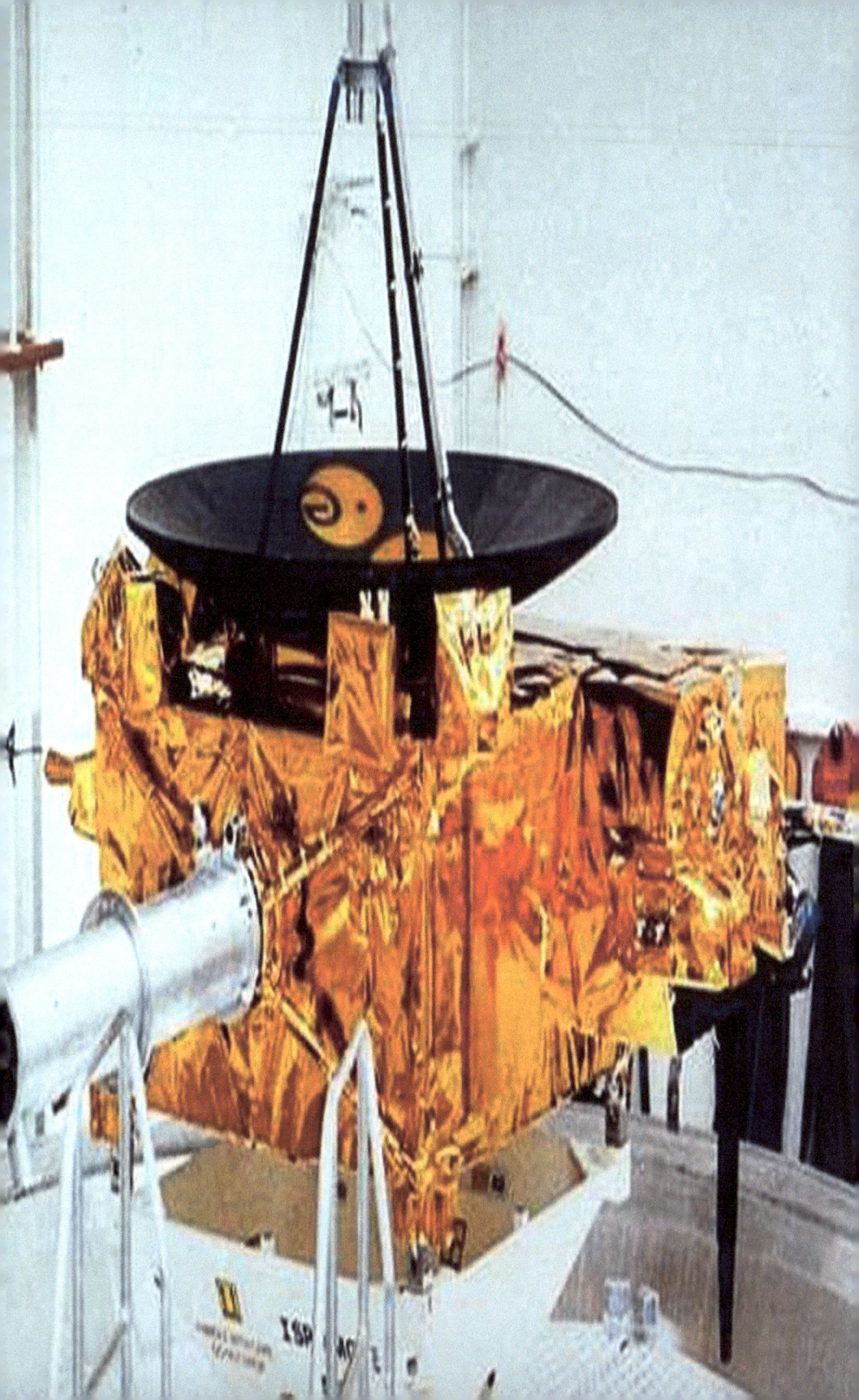

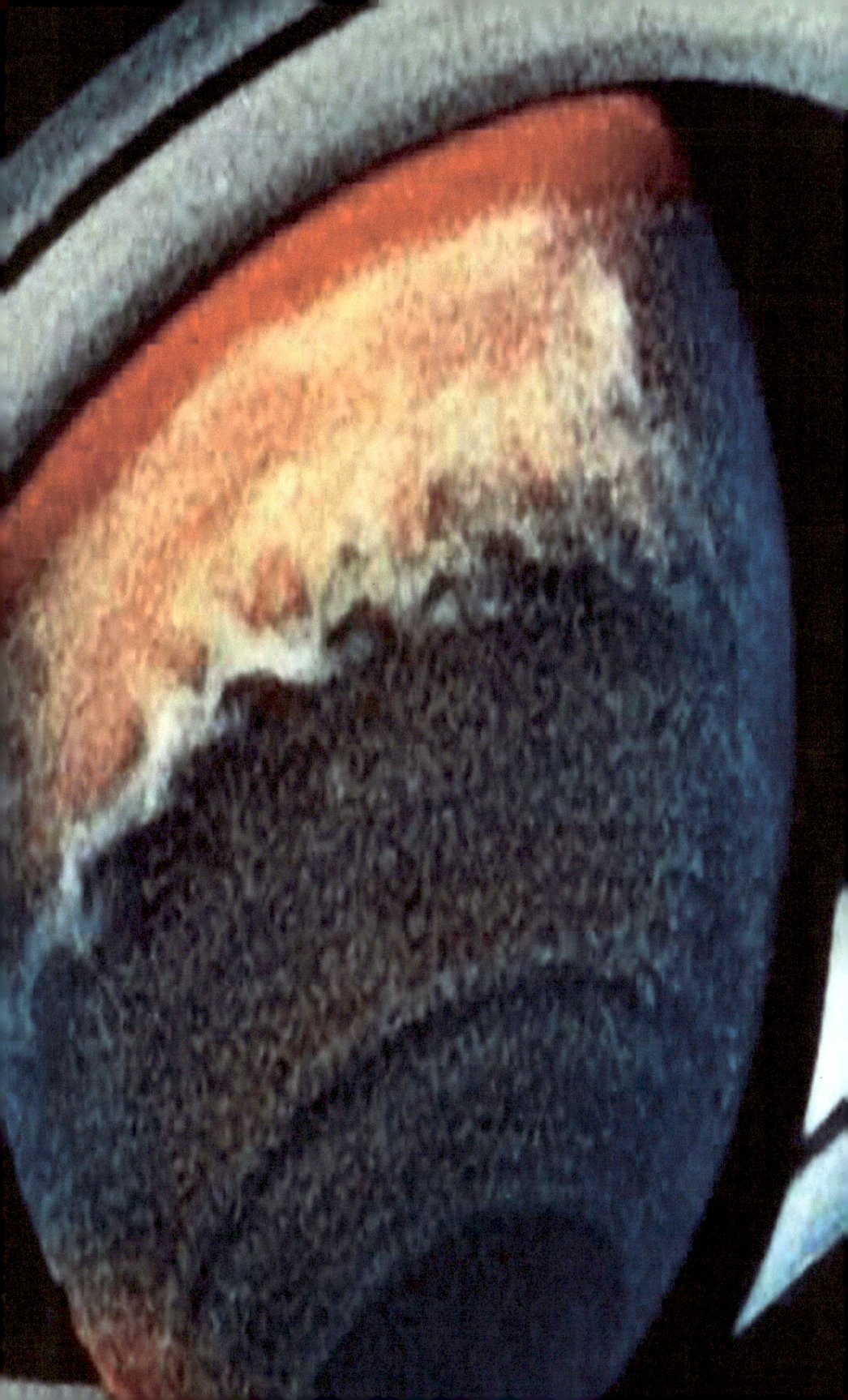

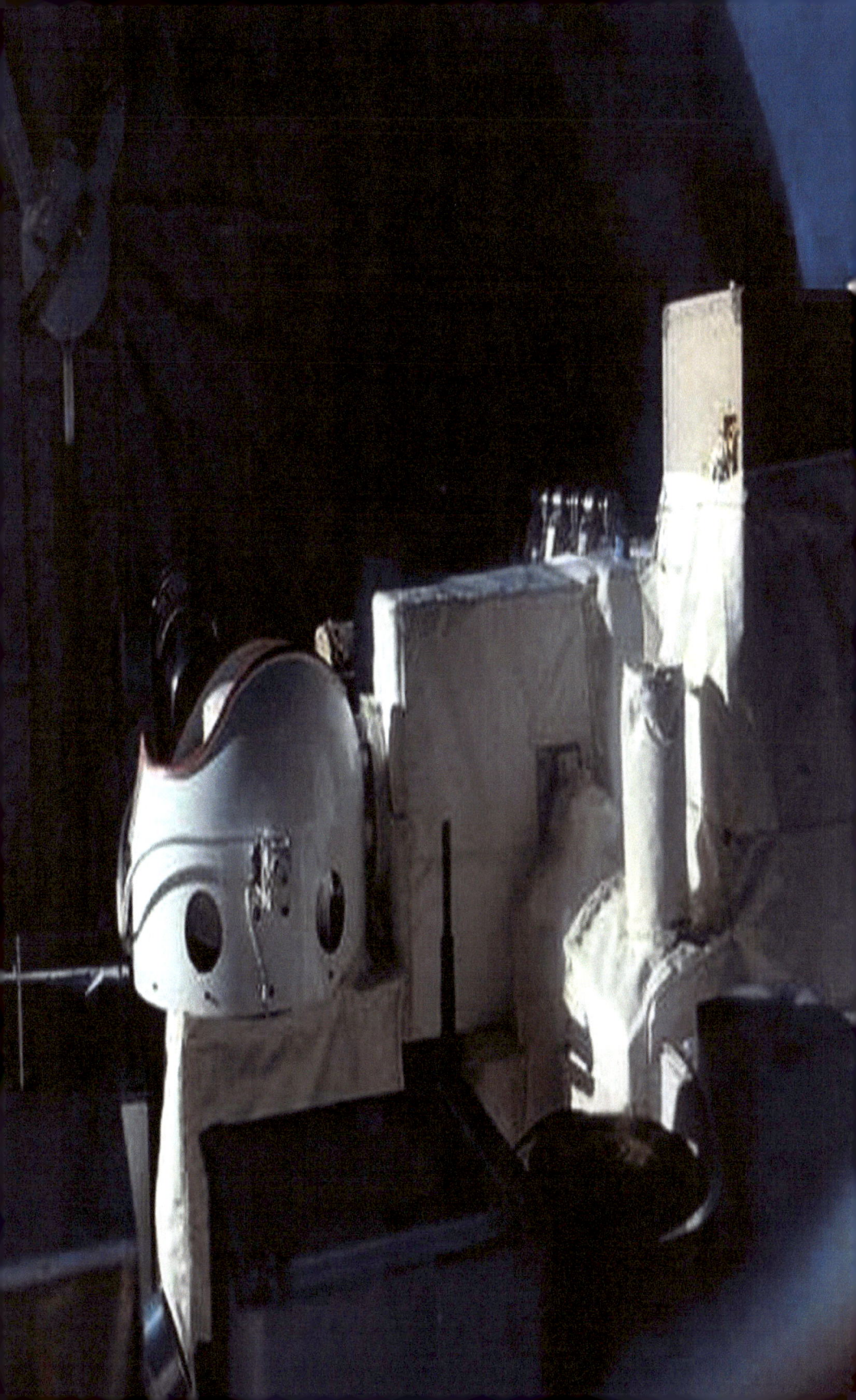

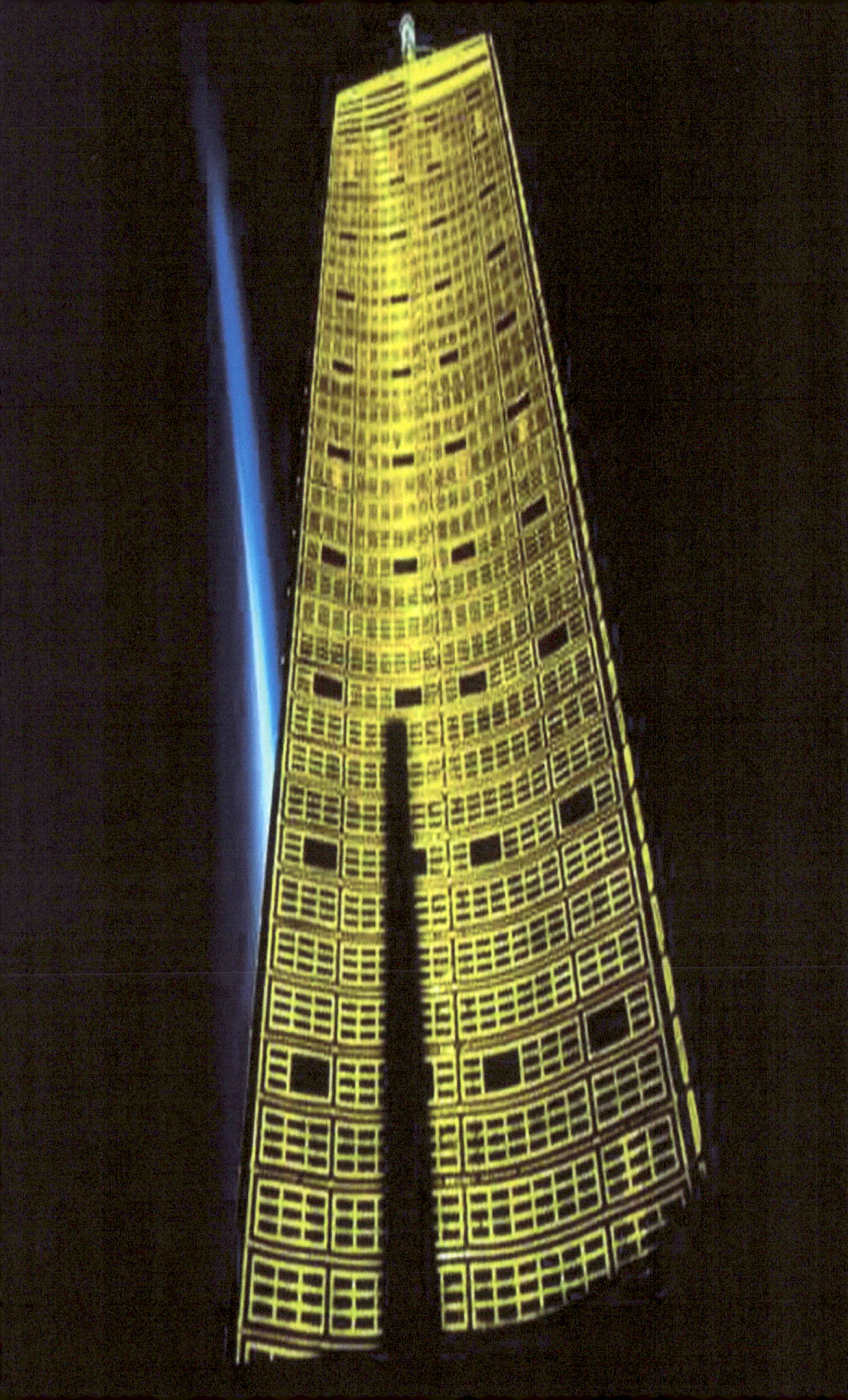

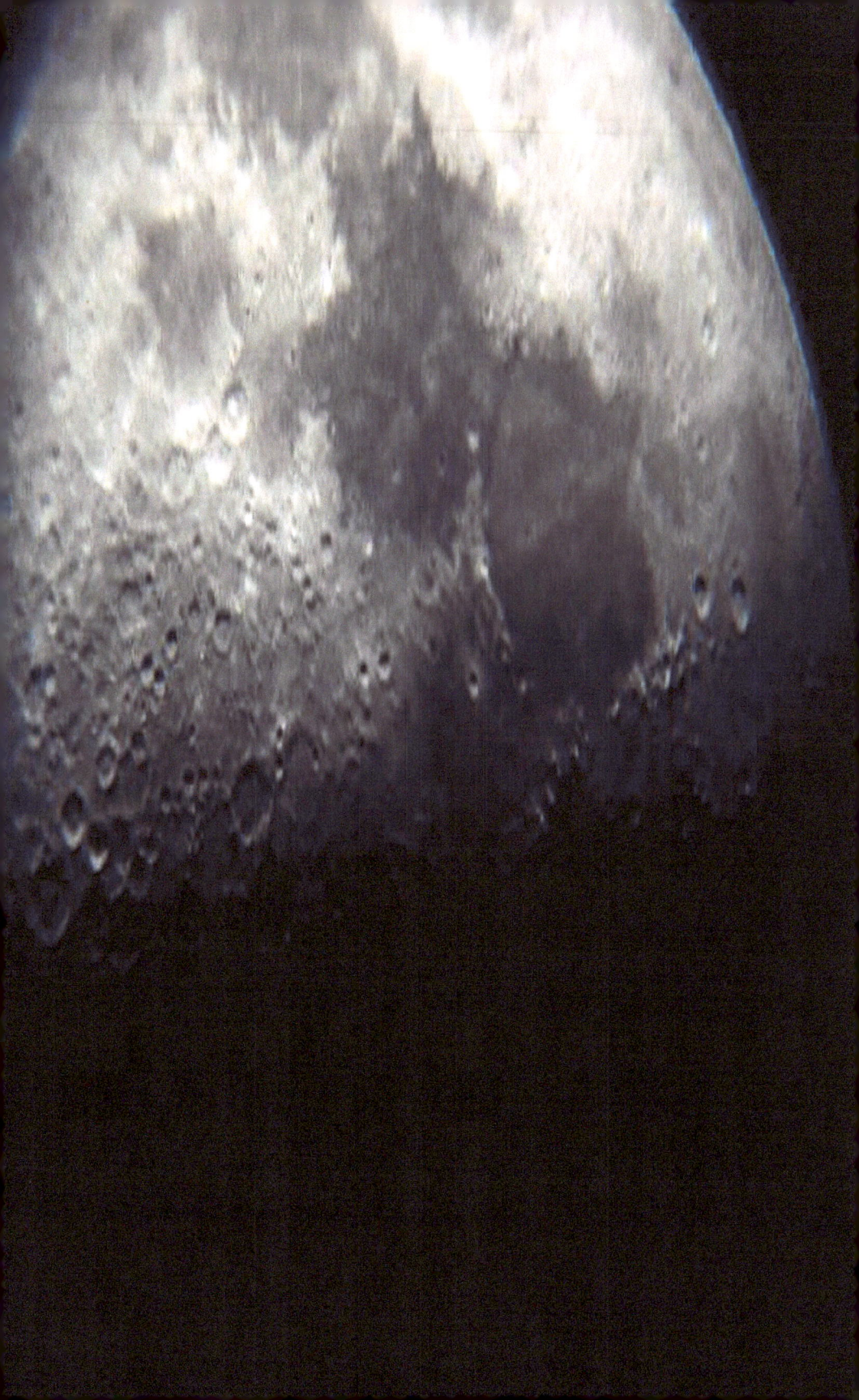

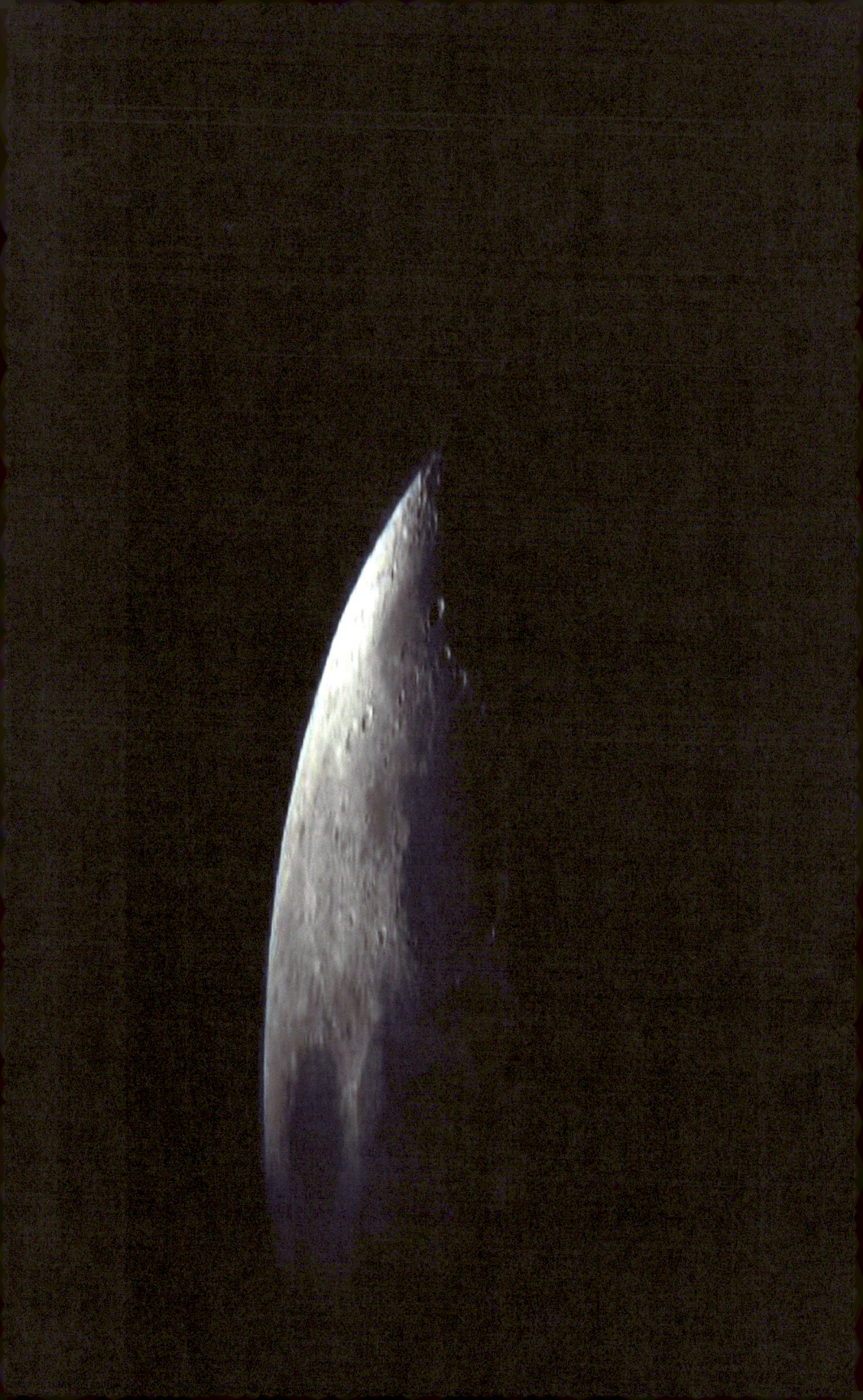

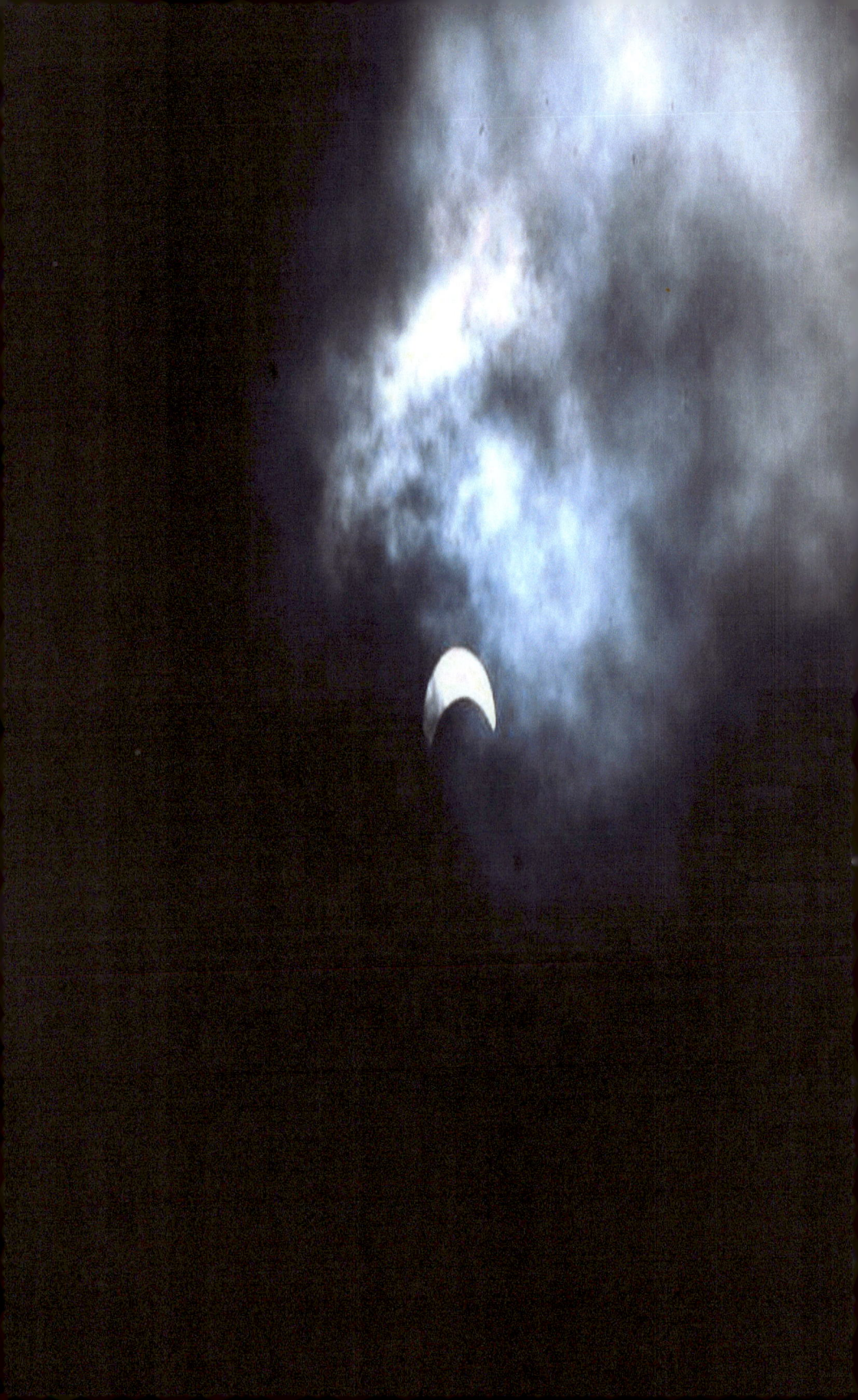

www.ingramcontent.com/pod-product-compliance
Lightning Source LLC
Chambersburg PA
CBHW041101180526
45172CB00001B/60